MIND-BLOWING MAMMALS

AMAZING ANIMALS

Larry Tackett, TOM STACK AND ASSOCIATES

SLOW LORIS
Nycticebus coucang

Leslee Elliott

Sterling Publishing Co., Inc. New York

With Love To Stefie
Who Fancies Furry Fellows

Library of Congress Cataloging-in-Publication Data

Elliott, Leslee.
    Mind-blowing mammals / Leslee Elliott.
       p.   cm. — (Amazing animals)
    Includes index.
    ISBN 0-8069-1270-7
    1. Mammals—Miscellanea—Juvenile literature. [1. Mammals—
Miscellanea.] I. Title. II. Series: Elliott, Leslee.   Amazing
animals.
    QL706.2.E44   1994
    599—dc20   94-26052
                                                  CIP
                                                  AC

Designed by Judy Morgan

Cover photo of gorilla by Joe McDonald, TOM STACK AND ASSOCIATES

1 3 5 7 9 10 8 6 4 2
Published by Sterling Publishing Company, Inc.
387 Park Avenue South, New York, N.Y. 10016
© 1994 by Leslee Elliott
Distributed in Canada by Sterling Publishing
% Canadian Manda Group, One Atlantic Avenue, Suite 105
Toronto, Ontario, Canada M6K 3E7
Distributed in Great Britain and Europe by Cassell PLC
Villiers House, 41/47 Strand, London WC2N 5JE, England
Distributed in Australia by Capricorn Link (Australia) Pty Ltd.
P.O. Box 6651, Baulkham Hills, Business Centre, NSW 2153, Australia
Printed and Bound in Hong Kong
All rights reserved
Sterling ISBN 0-8069-1270-7

Edited by Dr. Nancy B. Simmons, Assistant Curator, Department of Mammalogy, American Museum of Natural History

CONTENTS

CAMEL   *Dromedary—Camelus dromedrius*

MAM-MAL   a warm-blooded, air-breathing animal that feeds its young with milk, has a backbone, two pairs of limbs and usually a covering, more or less, of hair.

But that's not all!

There are actually three different kinds of mammals:

MONOTREMES	**MARSUPIALS**	**PLACENTALS**
3 species, including platypuses and spiny anteaters.	250 species, including kangaroos, koalas and opossums.	3,755 species, including dogs, horses and whales.
The most primitive animals that can still be called mammals, monotremes still have some things in common with reptiles. For example, they lay eggs.	*Born in a very immature state, marsupials finish developing in a pouch in their mother's belly.*	*Placentals develop to an advanced state before birth. Some can even stand and run within minutes of their arrival.*

AFRICAN ELEPHANT   *Loxodonta africans*

Joe McDonald, TOM STACK AND ASSOCIATES

SOME GIANTS ARE STILL HERE

In prehistoric times, giant animals—some of them two stories high—roamed the earth. There were huge camels and beavers the size of bears. In those days—from two million to 10,000 years ago—many animals were super-size! Even if the largest modern ELEPHANT weighed 22,000 pounds (10,000kg) and stood 13 feet (4m) at the shoulder, it was still only half the size of the Imperial Mammoth!

Now that the mammoths are gone, two kinds of elephants, the Asian and African, are the largest land animals. The African elephant is larger, with the world's biggest ears. But to tell the two types apart quickly, check out the trunk. Asian elephants have one lump on the end (called a lobe); African elephants have two.

Elephants have trunks for a very obvious reason. Their necks are too short to allow their mouths to reach the ground. Their trunk is an all-purpose tool that is used for just about everything: breathing, drawing up water for drinking, spray baths, maneuvering objects, greeting friends, hugging or spanking their offspring, even as a snorkel during deep-water crossings. But its most important use is gathering food.

WOULD YOU BELIEVE . . .

Elephants never stop growing during their entire 60-year lifespan, though they grow very slowly as adults. So the largest member of the herd is probably also the oldest.

Adult elephants can eat 330 pounds (150kg) of leaves, berries and twigs a day, all plucked with the trunk and placed in the mouth. Eating is very important to such large animals. They may wander 30 miles (48km) in one day searching for a good food source—often travelling down centuries-old paths called "elephant roads," worn down by generations of their ancestors. Once a suitable location is found, the day is spent in peaceful munching on tender leaves.

Plenty of food and a safe environment seems to make elephants happy, and like content kittens, they show their pleasure by purring. It's true—their tummies rumble with a sound that's a lot like a purr or low growl. The tummy rumbling noise carries for more than half a mile (1km) and serves to keep the elephants in touch with each other when they're in the dense brush. If one of them senses danger, its rumbling abruptly stops. The sudden silence warns the rest of the herd to be alert—kind of a reverse burglar alarm!

Thomas Kitchin, TOM STACK AND ASSOCIATES

ASIAN ELEPHANT   *Elaphas maximus*

CAPYBARA *Hydrochaeris hydrochaeris*

A MOUSE THE SIZE OF A HOUSE

. . . Dolls' house, that is. It's the South American CAPYBARA, at four feet (1.2m) long and 200 pounds (91kg), the world's largest rodent. "Chirping" along on partly webbed feet, the capy searches jungle-like areas around ponds for grasses, grains, water plants and fruit. Pretty quick on land, the capy runs just like a horse, but its first love is the water.

Water is a necessity for its good health (its skin is dry and gets sores without the moisture) as well as enjoyment. These rodents are as graceful as ballerinas in the water. They have "neutral buoyancy"—little or no weight in the water—and the slightest movement propels them across a pond.

Caimans (crocodilians) are the main enemy of the defenseless capy in the water. On land, avoiding jaguars and humans keeps the timid animal safe. Even though the capy isn't a threat to anyone, imagine your tame tabby's reaction to this giant mouse!

STEALTH, NOT SPEED

Would it surprise you to know that the shy, elusive LEOPARD is able to catch almost anything it feels like eating, from a tiny frog to an antelope larger than itself? Long considered Africa's cleverest hunters, leopards, are stalkers, not runners. They're also incredible jumpers—they can leap as high as a car length straight up and two luxury cars long! Quick as a wink, they're up in a tree, out of harm's way. This capable kitty, strong enough to lug a 50-pound (23kg) antelope dinner into the highest branches of a tree, is also smart enough to keep its kill away from other predators, such as lions and scavengers like hyenas.

Many leopards develop a taste for only one food—fish, for example. A leopard may live right next to a farm filled with all kinds of easy pickings, but never go near the farm animals, preferring to eat only its favorite food, even if it's much harder to come by. This partiality to a certain food may explain why some of the big cats became man-eaters!

AS AMAZING AS IT SOUNDS . . .

Sightseeing leopards raise their tails like a white flag, which lets other animals know they're not out hunting!

Jeff Foott, TOM STACK AND ASSOCIATES

LEOPARD *Panthera pardus*

9

CAMEL   *Bactrian—Camelus ferus*

ONE HUMP OR TWO?

Did you know that CAMELS originated in North America? Fossil records prove it. They show the smallest to have been about the size of a rabbit and the largest a giant of 15 feet (4.5m) at the shoulder. As the creatures multiplied, a few crossed land bridges to other continents. Some went to South America, where they survive as llamas. Others went to Asia and became two-hump Bactrian camels. From those, another camel descended—the Arabian (dromedary) camel, which had only one hump. It survives as today's one-hump dromedary, a special domesticated breed. Both camels live in the desert of the Middle East and are used for riding, but the Bactrian has shorter legs and is more easy-going.

The camel's ability to go for long periods without water is legendary. But just how long is long? When they're not working, camels can endure up to three months without water! When working or walking in the heat of the desert, they can go about a week with absolutely no moisture—or several weeks if there are water-filled desert plants to snack on.

BELIEVE IT OR NOT . . .

There is no "normal" body temperature for a camel, just a normal range. It goes from 83° to 106°F (28° to 41°C) depending on the weather. Why? To avoid sweating away precious water on hot days!

Water is not stored in the camel's hump. The bumps on its back are solid fat, an energy reserve. To survive with little water, the camel uses the moisture stored in its tissues. It can loose up to 25 percent of its body weight and still do a good day's work. In comparison, if you weighed 100 pounds

(45kg), you would be in big trouble if your weight dropped down to 75 pounds (33.75kg) from loss of water. Why? Because you would not only be losing moisture from your tissues, but also from your blood. Soon your blood would get thick and sticky and your heart would not be able to circulate it, and you'd be a goner. But take a look at the dehydrated camel, with its ribs sticking out and its skin hanging from its body. Ten minutes after that camel drinks 30 gallons (114 l) of water, it'll plump right up before your very eyes! Camels are the only animals in the world that can do this.

THE REALLY AMAZING PART . . .

A dried-out camel can lift the same 400-pound (182kg) pack and do the same work as a fully watered one!

11

WALK THIS WAY

In Central and South America, there is a creature that manages to walk gracefully and comfortably on its wrists! It's the ANTEATER, and its feet naturally turn under to protect the long claws it uses to rip open ant and termite hills—its favorite source of food.

You can guess from the anteater's very long nose that its sense of smell is good—40 times better than yours, as a matter of fact. It's good enough to sniff out the most hidden termite holes.

Surprisingly, in a face-off, the Giant Anteaster (3–4 feet/.9–1.2m) can usually fight off an attacker as powerful as a jaguar with its amazing coddled claws that are longer and sharper that those of a big cat! For the anteater, danger comes from above in the form of owls and hawks, its main enemies.

GIANT ANTEATER *Myrecophaga tridactyla*

John Cancalosi, TOM STACK AND ASSOCIATES

BANDED ANTEATER *Myrmecobius fasciatus*

MARSUPIAL WITHOUT A POUCH

Yes, this little creature is an ANTEATER, too, even though it's only about the size of a squirrel. "Banded" refers to its beautiful striped coat. A native of Australia, it usually goes by its Aboriginal name "numbat." Though most marsupials have a pouch, numbats don't. To get a ride on their mother, the tiny babies must cling to her long stomach hair. When the mother decides that they're too heavy to carry around, she stores them temporarily in a ground-hole while she looks for food.

Numbats like to sleep in hollow logs and these are their main means of defense. After entering the log and tucking their tail under, they plug up the opening with their rear end, swelling themselves up to fit in tight as a cork. Pythons and goannas (big monitor lizards), their major enemies, can't get a grip on them to drag them out!

FABULOUS FACT . . .

The name for "stuffing the log" is "phragmoticism"!

SPINY ANTEATER (ECHIDNA)   *Tachyglossus aculeatus*

NO ONE'S RELATIVE

This SPINY ANTEATER is no relation to either of the anteaters on the previous page. A native of Australia, it looks like a porcupine or hedgehog with a very long nose. But while it's a different species from the South American anteater, it has developed the same hunting equipment for catching the ants and termites it loves to eat: a long nose and a sticky eight-inch (20cm) long tongue. This is called "convergent evolution." It means that animals with different origins and ancestors end up with the same features because they live in similar conditions.

A monotreme, the spiny lays one single egg and carries it in a pouch. The amazing part is that the pouch does not exist until it is needed. Then the mother's flat belly muscles fold the stomach skin together to form a temporary egg-holder. After the baby hacks its way out of its shell, it stays in the holder for ten more weeks, lapping up the milk from the mother's pores. As soon as baby leaves, the pouch disappears, becoming ordinary stomach skin again!

WOULD YOU BELIEVE . . .

The numbat (page 13) is another example of convergent evolution. It's more closely related to the kangaroo than the spiny anteater!

WILL THE REAL UNICORN PLEASE STAND UP?

Did the unicorn ever exist? Well, we're pretty sure (aren't we) that it didn't, but if that's so, then where did the explorers find the prized unicorn horns that they brought back from their travels?

Let's look at the possibilities: Could the horn have come from the horned but ugly rhinoceros? No, that horn is too short. Or the male narwhal, a 16-foot (5m) member of the whale family, with its 8-foot (2.4m) long, twisted single tooth-tusk? No, too long and twisty.

DID YOU KNOW . . .

Treasured beyond gold because people believed it could neutralize poison, the unicorn horn was worth a king's ransom.

More likely the horn came from the Arabian ORYX, a pretty, smallish antelope-type of creature that is not often seen today but was once found throughout Arabia, Syria and Iraq. Unfortunately for the oryx, the horns were in such demand and the poachers so successful that the wild population became extinct in the early 1970s.

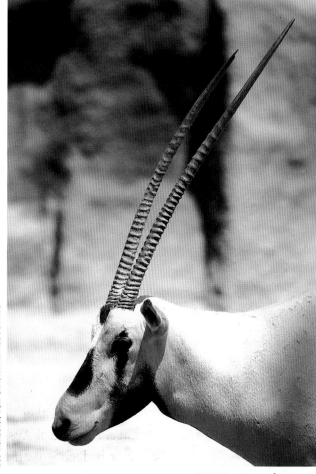

Bob McKeever, TOM STACK AND ASSOCIATES

ORYX   *Oryx leucoryx*

COPING WITH COLD

In the frozen wasteland of Alaska, the ARCTIC GROUND SQUIRREL sleeps nine months of the year. While the outside temperature is −50°F (−45°C), the squirrel's underground hideout is over 20°F (6.6°C), a full 70°F warmer! Safe from the killing cold, the little mammal practically suspends life for the winter. It goes into a very deep sleep. Its breathing slows; its bodily functions, such as heart rate and metabolism, slow down or stop altogether, and its body temperature drops to just above freezing. The hibernating squirrel only needs to use a little bit of energy to live through the winter on the fat stored in its body!

Those lazy nine months are going to cost it, though, because, when it does wake up, it only has three months left to do everything that it takes all year for a squirrel in a warmer climate to accomplish. Soon after waking, it mates and 25 days later its young is born. One month gone. The baby's eyes open three weeks after that, followed in three days by a search for roots and seeds. Now there's only one month left to grow, dig its own burrow and stock up on food for the coming nine-month hibernation. How does it do it? By working frantically at least 17 hours every day!

ARCTIC GROUND SQUIRREL
Spermophilus undulatus

HARRIS GROUND SQUIRREL *Ammospermophilus harrisii*

COPING WITH HEAT

The HARRIS GROUND SQUIRREL lives in the blistering heat of the deserts of the southwestern United States. The sand it walks on is sometimes 150°F (65°C). Its own body temperature goes up to 110°F (43°C). This hot-blooded rodent has a temperature every day, but it never gets sick! How does the squirrel do it?

For one thing, this squirrel doesn't lose any body water. In fact, even its urine is almost solid. When the heat gets to be too much, it dives down into its shady burrow. In the desert-burrow, the weather is cooler, because the soil insulates it from the heat aboveground. And if the fuzzy fellow has to go out, it can take an emergency measure and spread saliva all over its head to cool off. These cousins have really developed different ways to cope with the weather extremes they live in!

Joe McDonald, TOM STACK AND ASSOCIATES

CHEETAH *Acinonyx jubatus*

FASTER THAN A SPEEDING LOCOMOTIVE

No, not Superman, super CHEETAH! The fastest land animal is considered to be the hunting specialist of the short-grass plains. Usually unaggressive and mild-tempered, cheetahs need plenty of endurance to catch one of their favorite dinners—Thompson's gazelle. Called "tommies" for short, these gazelles are the most numerous animals in the plains community; herds of thousands of tommies bounce across the plain in an unusual gait known as "stotting," with legs stiff as a marionette's. Unfortunately for the tommies, they must also be quite tasty, because they are the prey of the jackal, hyena, Cape hunting dog and all the cats, including the cheetah, their biggest fan.

Cheetahs are built for speed, with their slim lithe

body, extra long legs and small head. They can sprint faster than most people drive—70 miles (113km) per hour—for short distances. They take advantage of these short bursts of fabulous speed with a specialized hunting technique. They approach the prey animal quite openly from afar, but always in profile, never head-on. Apparently, many animals do not regard the cheetah as a threat as long as it doesn't seem to be coming directly for them. When the crafty cat gets close enough, it quickly turns and charges, accelerating from 0 to 45 miles (72km) per hour in two seconds! Look out! But, it's too late for the tommie to escape!

DID YOU KNOW . . .

Wild cheetahs only like fresh meat that they've killed themselves. They never return to snack on leftovers!

Measuring the speed of an animal in the wild is very difficult to do. That's why, in the 1920s, in London it was decided to *scientifically* race the cheetah against the greyhound. The race turned out to be inconclusive. Many suspected the cheetahs were more interested in eating the dogs than racing against them. A safe piece of advice might be: Don't bother racing a cheetah. You can't win and you could become dinner!

Joe McDonald, TOM STACK AND ASSOCIATES

SQUIRREL MONKEY   *Saimiri sciureus*

MONKEY BUSINESS

If you think of apes as "first cousins" to humans, then MONKEYS would be second cousins. Apes, chimps and gorillas have a body structure that is closer to our own, with a generally heavier build, and a broad chest. Their organs, muscles and bones are also more suited to an upward stance. Monkeys, on the other hand (except for ground dwellers like baboons), usually have a longish, narrow body. Their organs are grouped like other four-footed animals. And, of course, monkeys have tails and apes don't. But maybe you've already guessed the most important difference. Apes far outrank monkeys in "brain power." More about apes later.

Monkeys are divided into two groups by geography. New World monkeys—the cebids and marmosets—live today in tropical Central and South America. Old World monkeys—langurs, baboons and others—live in Africa and Asia. Separated from each other for at least 30 million years, each group has developed its own way to cope with its environment.

IN FACT . . .

The most "sociable" award goes to the squirrel monkeys. They are always chattering.

Most striking is the "fifth hand," the "prehensile" tail of some of the New World monkeys. This fabulous tail is delicately ridged almost like a human finger and it can not only grasp a tree limb, but it also is capable of picking up a small peanut! Three out of six species have this feature: the cebus, the howler and the spider.

Another interesting feature: tough rump callouses, called ischial callosities, have developed in the Old World monkeys. These callouses allow the creature to sleep comfortably on a branch in a sitting position.

WOULD YOU BELIEVE . . .

A gibbon can touch its toes without bending over!

New World monkeys regularly have twins and triplets and the father is likely to take part in the child rearing. Old World monkeys usually have just one offspring and its mother will take care of it.

A simple way to tell New World and Old World monkeys apart is to look at their faces. A broad nose with well-separated, round nostrils that face outward—almost—belongs to a New World resident. A narrow nose with downward-pointing, comma-shaped nostrils belongs to an Old World monkey.

Kevin Schafer/Martha Hill, TOM STACK AND ASSOCIATES

MANDRILL   *Mandrillus sphinx*

DID YOU KNOW . . .

Mandrills are the most brilliantly colored baboons. Their faces flush even more intensely just before a fight. Sometimes that's enough to scare the enemy away.

WOULD YOU BELIEVE . . .

A legend from South African bushmen claims that baboons can talk. But they never let men hear them for fear of being put to work!

BEST OF BOTH WORLDS

BABOONS are some of the largest Old World monkeys (see page 21). Males can reach 100 pounds (45kg), females half of that. Baboons take advantage of the best of both primate worlds. They don't live exclusively in trees (like most monkeys) or on the ground (like most apes). They search the ground for easy-to-find food during the day and escape the reach of predators by sleeping in the trees at night.

Baboons live in big families called troops. The arrangement of a troop moving across the open grassland is very specific. Smaller males walk to the front and rear; older dominant males cover the sides; females and youngsters travel within this caravan, and at the center come the mothers and babies. If any monkey spies danger, it gives a sharp bark. A male troop member will immediately move away from the group to watch the intruder. If the intruder makes the slightest move, a double bark sounds the alarm. All the youngsters run for their mothers; the females carrying infants escape first, then the juveniles, the females and finally the small males. The dominant males are likely to turn and stand their ground. They are very brave and actually capable of fighting off a leopard!

MAY I BORROW YOUR COMB?

Wouldn't it be great never to lose your comb? The SLOW LORIS doesn't ever have that problem—because its comb is in its mouth! Its evenly spaced teeth get rid of tangles, while under the loris' tongue is a feather-like kind of finger called a sublingua that removes the bits of fur that get stuck in this dental comb!

INCIDENTALLY . . .

Being slow also makes sneaking up on your dinner easier. Surprisingly, the loris is able to grab its prey with amazing speed!

The loris is about the size of a house cat with the shape of a monkey (it's related), and covered with soft, fluffy hair. And this hair is very important, because the loris has such a slow rate of **metabolism** that losing its hair—even in the tropics, where it lives—would mean dying of the cold!

So how does such a slowpoke escape from the snakes, cats and birds that want it for lunch? Apparently, pretty well! When startled, the loris simply lets go of the tree branch it's hanging from and falls out of reach. It's risky, but the only means of a quick exit for the loris.

Rod Planck, TOM STACK AND ASSOCIATES

SLOW LORIS   *Nycticebus coucang*

DID YOU KNOW . . .

Getting bitten by the sweet-looking loris can be risky for humans. Its toxic saliva can cause a severe allergic reaction—even death!

OTTER   *Lontra canadensis*

YOU WOULDN'T KNOW A MONSTER IF YOU SAW ONE

You've heard of the Loch Ness Monster and maybe you've seen pictures of it—you know, sea-serpents drawn with a head and a line of humps following? There are supposed to be lots of these lake monsters. Well, they may be easier to explain than you think! The long sleek OTTER naturally swims showing three humps: the head, back and tail. After their mothers give them their first nudge into the water, the young otters swim with her every day, following her lead in a perfect line of little humps. Sometimes two families swim together. When the mother in the lead lifts her head to see "What's happening?" Bingo! There are the head and humps of the monster!

Otters like the cold water. Some of them even hang out around Arctic ice! They're comfortable because of the 800 million fur fibres that trap warm air next to their body. These fibres are so dense that water never touches their skin.

Otters don't hibernate. They keep fishing even when the lake is iced over, diving down and coming up to breathe through holes in the ice. If there aren't any convenient breathing holes, they come right up under the ice and breathe out. This creates an air-bubble that loses carbon dioxide into the ice and water. The oxygen trapped in the ice rushes in to take place of the carbon dioxide. The otters then inhale the oxygen that is now in the bubble and go about their business!

CAN'T CATCH ME!

Zooming across the desert at a blistering 35 miles (56km) per hour, the BLACKTAIL JACKRABBIT zig-zags like a professional football player. Usually, it outruns its enemies, but if that doesn't work, it flashes a quick change, turning its body towards the enemy very slightly, so that its hair appears to change color from its normal tan to white. This ability is called "directive coloration," and it confuses the pursuer long enough for the rabbit to escape.

Hares also count on speeding to safety. Their favorite trick is called "jinking." Just when it looks as if they're about to be caught, they turn sharply to the left or right at full speed, sending the hunter racing straight ahead after empty space. Watch the hare's ears! The hare always flattens its ears against its head just before jinking.

DO YOU KNOW . . .

How rabbits are different from hares? In addition to being a little smaller, true rabbits give birth to many naked, blind young in underground nests. Baby hares are born above-ground in the open, one or two at at a time, with a full coat of hair, working eyes—and in less than five minutes they are able to run!

Wendy Shattil/Bob Rozinski, TOM STACK AND ASSOCIATES

BLACKTAIL JACKRABBIT *Lepus californius*

THE FURRIEST ANIMALS

In the cruel environment of northern Canada, Alaska and Greenland, where the temperature hovers at a breathtaking −50°F (−45°C), and storms frequently drop that by another mind-boggling 20°F, survives the animal that the Arctic natives call *oomingmak*, or "bearded one." Very distant relatives of wild cattle, MUSK OXEN look like sloppy water buffaloes. The "big bruisers" are the furriest animals alive! They have a six-inch (15.25cm) thick layer of dense, wooly hair called "quiviut," and guard-hairs so long that they brush the ground. This extraordinary coat helped their ancestors survive the Ice Age, and is even protection from Eskimo arrows.

AS A MATTER OF FACT . . .

Musk oxen are so well insulated that when they lie down, the snow does not melt under their bodies!

Today musk oxen are protected from humans by the Canadian government. The only thing they need to fear is hungry wolves. No problem. Musk oxen stick together when the wolves start to prowl, forming a tight ring—a "circle-the-wagon" maneuver—with their tail ends to the inside and their sharp horns pointed out towards the enemy. The young are safely sheltered in the center as the big guys make lightning-quick strikes at the wolves, trying to gore them. This strategy holds off even the most determined wolf!

Brian Parker, TOM STACK AND ASSOCIATES

MUSK OXEN *Ovibos moschatus*

RATS   *Rattus villosissimus*

RATS!

Since the first man climbed on a raft, seamen have been plagued by RATS. As stowaways on every oceangoing vessel, these resourceful, fuzzy creatures have also been the ultimate carriers of death. It was tiny fleas living in the fur of rats that spread the Black Death, a plague that killed half the people in Europe during the Middle Ages. But the rats managed to stay alive! Like the other animals you're reading about in this book, they have many strategies for survival.

For example, on the other side of the world in the Pacific Ocean are hundreds of surface-level coral islands and reefs that are next to impossible to see from the deck of a ship. Naturally, before radar, ships often ran aground and were destroyed. A scientist spending the night on a barren (that means no trees, no name, no nothing) coral island, couldn't figure out how the hundreds of shipwrecked rats living there had managed to stay alive! No food, you see. Then one day he discovered the rats fishing, or more correctly, crabbing! These clever rodents would sit on the rocks with their tails dangling in the shallow water, waiting for a crab to foolishly take the "tail bait." The rats got fed, the crabs—well, never mind about that part.

DIGGING THE DIRT

The one and only member of the order *Tubulidentata*—the AARDVARK—lives only in Africa. The most unusual and specialized of the termite-eaters has a remarkable talent—it's a spectacular digger! Aardvarks can dig out a four-foot (1.2m) long burrow faster than a team of six men digging furiously with shovels. The aardvark's powerful claws make it easy for it to get into the rock-hard termite mounds to devour its favorite dinner. Its tough, bristly nose skin—and the dense nose hairs that close off its nostrils during digging—keep the aardvark safe from termite bites, while its 18-inch (45cm) long tongue scoops up the insects—ants in the dry season, termites when it's wet. The surface of the aardvark's teeth is like no other mammal's. There's no enamel on them; instead they are covered with a substance called cementum, which is usually found on the roots of other animal's teeth. Cementum is made of mineral salts and water, and is about as hard as bone.

WOULD YOU BELIEVE . . .

Aardvarks sometimes smell like rotten fruit, which is lucky for them because that smell attracts their favorite dinner—insects!

AARDVARK *Orycteropis afer*

Gary Milburn, TOM STACK AND ASSOCIATES

DOLPHIN *Delphinus delphis*

GOOD SAMARITAN OF THE SEA

Dolphins are mammals that long ago renounced their land-dwelling ways and returned to the sea. They are actually small whales that can swim better than some fish. The flippers that guide them through the water are all that remain of what were once front limbs. Traces of bony skeleton indicating long-gone legs are now covered by a sleek swim-mer's body. And though air is what they breathe, they spend 98 percent of their time underwater.

Even though you can't see their ears, dolphins have super hearing. Their use of **sonar** was discovered by accident during World War II. Arthur McBride, head of the world's first dolphinarium, in Florida, was trying to catch a bottle-nosed dolphin in murky water without success when he realized that the dolphins were making very rapid clicking sounds as they swam at breakneck speed towards the waiting net traps. Then, at the last minute, they veered off. The dolphins "knew" the nets were there without being able to see them! Dolphins use this ability—called **echolocation**—as a kind of super-sense, to avoid running into danger, to locate food, even to communicate. Here's how it works: The dolphin sends out whistles or clicks at the rate of 20 to 800 per second (far beyond what humans can hear). The speed and the way the sound bounces off an object and returns to the dolphin gives the animal a wealth of information about the objects nearby—their size, speed, shape, direction, and whether they're food or non-food.

Communication is very important to this species that depends on cooperation for gathering food. A small group of dolphins may start a "fish roundup" by swimming 30 feet (9m) apart, herding the prey towards the surface. They take turns breaching—jumping out of the water and slapping back in—stunning the fish and causing them to panic. From as far as six miles (10km) away, other dolphins will hear the activity and join in. Twenty or 30 individuals may grow to an efficient fish-catching group of 300 dolphins working together.

Dolphins not only communicate and work together, they help others. In November 1993, a dwarf pygmy whale was spotted heading for Tigertail Beach, Marco Island, Florida. Accompanying it were two dolphins, one on each side of the floundering and obviously ill whale. Witnesses insist the dolphins were helping the whale by holding the sick animal upright on its way to beach itself on the shore. They stayed with the sick whale until it got all the way up onto the sand. Then they headed back out to sea.

Brian Parker, TOM STACK AND ASSOCIATES

OKAPI *Okapie johnstoni*

AN ANIMAL WITH A 4-PART STOMACH

It's not surprising that a secretive animal from the dense rainforests of the eastern Congo was one of the last large mammals to be discovered (1900). The solitary OKAPI is the giraffe's only living relative. Although the okapi is much smaller (only about as tall as a man), its body shape is similar to a giraffe's and it was thought at first to be a cross between a giraffe and a zebra!

Okapis, along with cows, camels and many other mammals that have hooves instead of claws, are called ungulates. Many of them also chew a "cud." Scientists refer to them as ruminants. All ruminants are plant-eaters. Plants are great sources of carbohydrates (sugars and starches), but usually very low in protein, which is a must for growth and cell repair. This means that the okapi must utilize every bit of protein in the plants it eats. Nature has solved this problem by developing a complex digestive system for the ruminants, which is like no other. It is designed to break down the tough plant fibres that would otherwise be indigestible.

DID YOU KNOW . . .

Okapis' tongues are black, giraffes' blue!

This is how it works: When the animal takes a bite of food, it doesn't bother chewing too much and quickly swallows it into the first stomach (rumen). After digesting for a while and, when the okapi isn't busy eating new food, the stomach contents are returned to the animal's mouth to be chewed at leisure (this is the "cud"). Finally, the food goes on to *three* additional stomachs where microorganisms process and reprocess it until every nutrient possible is extracted and used! Of course, it takes a long time—almost four days—to completely digest one meal! But don't get the idea that okapis eat only once every four days. New food is eaten all the time. The stomach can tell the difference between digested and undigested food and moves it along the "digestion track" at the right time!

INCREDIBLY . . .

The okapi neck is very flexible, able to move every-which-way. Combine that with an extremely long tongue and you've got an animal that can lick itself anywhere and everywhere!

Roy Toft, TOM STACK AND ASSOCIATES

WORLD'S TALLEST ANIMAL

From the point of view of the tallest animal in the world, the GIRAFFE, the okapi is a midget. Look at it this way: Your dad can stand eyeball to eyeball with the okapi, but it would take your mother on your dad's shoulders, plus your sister standing on your mother's shoulders and you on top of the acrobatic tower to see eye-to-eye with the tree-topping giraffe!

DID YOU KNOW . . .

Giraffes live in the tree-dotted grasslands south of the Sahara Desert in Africa.

It's hard to believe that those long necks have the same number of vertebrae as ours—seven. Each vertebra is drawn out to a length of eight inches (20cm) or more! As holder of the "World's Tallest Animal" record, giraffes use their height and keen eyesight to keep tabs on each other from as far away as a mile (1.6km). You can tell a male from a female giraffe from quite a distance by the way it eats. Females bend their heads down over their meal of leaves and twigs, while the males stretch up to their full height, trying to reach the highest new shoots. This curious habit reduces the chance of competition between the sexes.

Giraffe mothers usually give birth at the same

Barbara von Hoffmann, TOM STACK AND ASSOCIATES

GIRAFFE *Giraffa camelopardalis*

HARD TO BELIEVE, BUT . . .

Adult giraffes sleep little, if at all—about a half hour in every 24, in 5-minute catnaps!

"birthing grounds," probably where they themselves were born. Although their disposition is pretty "laid back," they're very protective of the 12 calves that will be born to them in their lifetime, and always on the lookout for any lion foolish enough to try to run off with a baby. Their heavy soup-plate-size hooves can easily bash in the skull of the biggest cat!

Giraffes appear neat and clean, because, unlike other huge animals such as the elephant and rhinoc- eros, they don't go in for mudbaths. What they do like is a bite of salty-tasting dirt, now and then, because it contains minerals their body needs.

AMAZINGLY . . .

Startled giraffes can gallop 30 miles (48km) per hour for short distances.

Brian Parker, TOM STACK AND ASSOCIATES

WESTERN CHIPMUNK *Tamius minimus*

SAVED FOR A RAINY DAY

We all know that some rodents gather food to save for the time of year when no food is available. We call this behavior "hoarding." People do it all the time (go look in your closet). but the first-place hoarding award absolutely goes to the CHIPMUNK. Up to a bushel of nuts (that's enough space for 30 litres, eight gallons or 128 eight-ounce glasses of milk!) may be stored in its main burrow, just where you'd expect to find them—under the bed! It's true! Chipmunks prefer to sleep *on* their stash! And that's not all. This industrious storer has small emergency stashes hidden all over its territory. Sometimes it forgets just where it left some of the buried nuts. But that's part of a clever plot by Mother Nature. In the spring the forgotten seeds will germinate and new trees will grow, providing a continuing food source for our furry friend.

THE TRUTH IS . . .

Chipmunks and squirrels don't "remember" where they bury nuts, but they can smell them—even under a foot of snow!

SHORT-TAILED SHREW *Blarina brevicauda*

SURPRISING SHREWS

Dolphins (pages 30–31) and bats aren't the only animals that use *echolocation* for communication—so does the SHORT-TAILED SHREW! And not only that! For their size (4–7 inches/10–19cm long), these small, fuzzy ancestors of the first true mammals may also be the fiercest creatures on earth! And this is despite the fact that they are born blind, deaf, without hair or teeth and some no bigger than a peanut.

BY THE WAY . . .

If you should ever come upon a shrew, don't pick it up. There are only a few venomous mammals and this is one of them!

ON THE OTHER HAND . . .

Not all shrews are poisonous, but how will you know the difference?

The "Atlas" of the shrew clan is the four-inch (10cm)—four more inches are tail—HERO SHREW. This tiny animal can support a medium-sized man on its back! How does it keep from getting squished? The hero shrew's arched backbone is the most unusual in the animal kingdom. It is several times thicker and stronger than the shrew's other bones, and the backbones of other shrews. Its strength comes from its extra width and from ridges that lock one vertebra to the next. What use is this feature to the shrew? Only Mother Nature knows for sure!

CARIBOU *Rangifer terandus*

KEEPING WARM

The North American CARIBOU are the larger, wild version of Europe's reindeer. They roam the frozen Canadian plains, coming together in herds of thousands to make the annual migration to summer feeding grounds. They often stop along the way to paw a hole in the snow, hoping to get at their favorite meal, a lichen that grows in the far north called "Caribou moss." Lucky for the caribou, there's no competition for this sparse food. Its acid content makes other animals sick. Not the caribou, which thrives on 12 pounds (5.4kg) of the quick-burning, carbohydrate-rich moss every day. This heat-producing food-fuel keeps the caribou's body temperature at a comfortable 105°F (40.55°C).

To prevent heat loss, those long skinny legs have a temperature that is 50°F (27°C) cooler than the body!

Just as your clothes keep you warm, the caribou's hair keeps it warm. The club-shaped hair is thicker at the outside tips than at the base, forming a thick outer layer that traps tiny warm-air spaces close to the skin, with a fine curly underwool. The coat is so warm that the caribou seems completely unaffected by the cold weather.

BY THE WAY . . .

**A caribou will never be able to sneak up on you.
Its ankles click when it walks!**

To make the treacherous migration over slippery ice and snow, its large feet work like snowshoes. The two halves of the cloven hooves are flattened out, which reduces their pressure on the ground to two pounds (.9kg) per square inch, a very light weight. An animal of similar size, the moose, which doesn't have flattened feet, exerts eight pounds (3.6kg) per square inch. Without their big feet, the 700-pound (318kg) caribou would sink like a rock in the soft snow!

Thomas Kitchin, TOM STACK AND ASSOCIATES

HIPPOPOTAMUS   *Hippopotamus amphibus*

MANNERS MATTER

Some animals have what certainly look like "rules of social behavior" that the whole group respects and abides by. Consider the HIPPOPOTAMUS: The Greeks called it the River Horse, and with good reason. The huge muscled creature likes to spend most of the day in the water, floating and snoozing, sometimes submerging for a half hour. Actually, it has to stay wet. The water keeps its delicate skin from burning. The hippo's unique skin structure allows water to evaporate five times faster than it does on our skin. To protect itself, the mighty mammal even has its own built-in suntan lotion, a pink substance produced by glands under the skin that prevents the burning ultraviolet rays from getting through. This fluid is so red that many natives insist the hippo "sweats blood"!

ACTUALLY . . .

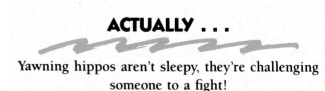

Yawning hippos aren't sleepy, they're challenging someone to a fight!

Groups of 20 to 100 hippos live together. The children and females stay in a central area called a "crèche." This is close to but separate from the adult males, each of whom has his own separate "refuge." When it's breeding season, the female chooses her mate by entering his refuge. A male may visit a female too, but only with her approval. When he enters the female area, the male must be calm and never aggressive. If one of the females gets up, he must lie down and not get up until she lies down again. Failure to follow the "rules" will result in the male being attacked by all the females.

AS A MATTER OF FACT . . .

Mother hippos are even happy to "spell" each other by babysitting!

Hippo society is ruled by the females and the males abide by their wishes. Hippo mothers teach their young the "rules" of good hippo manners, including how to follow instructions and pay respect to their elders. When walking, the youngster must remain exactly at its mother's shoulder. If she starts to run, so must baby. If she stops, baby stops. She teaches her youngster everything it needs to know to get along in the hippo world!

BOOMERS AND FLYERS

The world's largest marsupial—the GREY KAN-GAROO—can grow to 8 feet (2.4m) tall and weigh over 200 pounds (91kg). On most continents, deer, antelope, goats and horses eat grass and vegetation that grows on the open plains. In Australia's vast, arid interior, the kangaroo has taken the place of these familiar four-hoofed grazing animals. It spends the day sheltering from the sun's heat and the night peacefully feeding. The kangaroo's biggest problem isn't animal predators, it's the human race. Not only do people kill kangaroos but they also brought the sheep to Australia (it's not a native species) that compete for the same grass! Fortunately for the "boomers" (male kangeroos; females are called "flyers"), their digestive systems are very efficient. A part of the kangaroo's stomach uses bacterial action to break down the hard-to-digest plant cellulose that it eats. So, even though the kangaroo is much larger than the sheep, it only has to eat about the same amount to produce a far superior result—kangaroos are 52 percent muscle, sheep only 27 percent. During the dry season, when food and water are in short supply, kangaroos use their strong front limbs to dig for and create water holes that many different animals use—even where no surface water is visible.

Kangaroos frequently have to travel long distances to get a meal, and here's where the animals' amazing speed comes in handy! Kangaroos travel in great leaps of up to 27 feet (8.2m) long and 10 feet

GREY KANGAROO *Macropus giganteus*

DID YOU KNOW . . .

To be called a kangaroo, the animal's feet must be longer than 10 inches (25cm). Wallabies' feet are shorter!

BELIEVE IT OR NOT . . .

Nervous kangaroos lick their forearms!

(3m) high, at 20 miles (32km) per hour for distances, with bursts of 55 (88km)! While hopping may look like an awkward means of movement, it's much more energy-efficient than four-legged running. The kangaroo's strong leg muscles and huge feet are the power behind the jump and how it gets its family name, *Macropodidae* (*macropus* means big feet). The large tail acts like a counterbalance at

landing. As you can imagine, those big feet also make formidable weapons! They're not only gigantic, but equipped with a sharp claw that can be used to lash out and down in one of the most powerful blows that can be delivered by any mammal! The usually docile kangaroo can quickly kick almost any enemy to death with those fantastic feet!

NOT SURPRISINGLY . . .

Only the males get into fights. The flyer saves herself for motherhood!

Dave Watts, TOM STACK AND ASSOCIATES

CONVEYER BELT TEETH

Looking like small (2–3 feet/.6–.9m tall), very furry kangaroos, the ROCK WALLABIES of Australia are also marsupials. Like all marsupials, their babies are born after a very short time and finish developing in their mother's pouch. They're called "Joeys." The little rock wallaby is different from all its relatives. For one thing, it's only half their size, but the most important difference is its teeth. When you lost your baby teeth, another tooth was waiting below each one, ready to push right up through the gum and fill the empty space. The little wallaby's nine molars stand in a row waiting for the forward one to get worn down. When it falls out, the next in line moves forward to take its place. Only two other mammals have teeth like this—the elephant and the manatee!

BELIEVE IT OR NOT . . .

Some burrowing marsupials (wombats and koalas) have upside-down pouches. This is a convenient arrangement because it means that dirt and rocks don't get inside!

ROCK WALLABY *Petrogale inornata*

John Cancalosi, TOM STACK AND ASSOCIATES

LION *Panthera leo*

FAMILY PRIDE

LIONS are the only sociable wild cats. Their families are known as "prides" and are usually made up of 20 to 30 animals. One dominant male, several lesser males and many females and cubs cooperate in the chores of hunting and defending the pride.

Even though lions can run 40 miles (64km) per hour for short periods, jump up one story and leap over four car lengths, they usually don't climb trees like many of the other cats do. Lions rely on surprise, cunning, silence and, most of all, teamwork, for a successful hunt. They are master strategists, seeming to confer with one another to devise a successful plan.

For example, let's say the pride has spotted a small group of wildebeests and carefully positioned the main ambush group downwind, hidden in the grass. Two smaller groups of lions break off and circle around in opposite directions. coming up to the side and behind the wildebeests. Their plan is to drive the prey straight into the jaws of the waiting ambush party! It works—and dinner is served!

45

HOW COLD IS COLD?

To avoid the intense Arctic cold—as low as −70°F (−55°C) with a 30-mile (48km) per hour wind—most animals hibernate or migrate. Yet, one of the largest carnivores, as well as the largest of the bears, does neither. POLAR BEARS choose to live on the permanently frozen landscape year round. Scientists estimate that about 15,000 polar bears roam the 5 million square miles (13,000,000 sq km) of land around the North Pole. That's not really many bears, only one for every 333 square miles (866 sq km). In some areas, the local people, called Inuits, claim that they have never seen the "wandering one" (polar bear). They call it the *pihoqahiaq.*

WOULD YOU BELIEVE . . .

Seals in the Antarctic are much calmer than their Arctic cousins, probably because there are no polar bears at the South Pole!

How does this amazing animal cope with the incredibly cold weather? For starters, the bear is warm-blooded. In other words, its internal thermostat keeps its body temperature at a steady 100.8F (38.22C), but that's not the only thing. Its body shape is designed to retain warmth. It is rounded and compact—no big ears or long tails (big heat losers). And since large objects lose heat more slowly than small ones, the bear, which can weigh up to 1,600 pounds (726kg), stays warmer longer!

Jeff Foott, TOM STACK AND ASSOCIATES

POLAR BEAR *Thalarctos maritimus*

To maintain its size and all those layers of fat, the polar bear often eats 100 pounds (45kg) of its favorite food, seal blubber, at one sitting. Half of the food it eats is used to keep up its body temperature (same as for humans). The colder it gets, the more food the animal needs to keep warm. Naturally, polar bears eat lots!

AMAZING BUT TRUE . . .

Polar bears can smell a dead meal 20 miles away and a live seal three feet (1m) under the ice!

AND BESIDES THAT . . .

Legends claim that the white bear covers its black nose with its paw when stalking prey!

Polar bears don't mind jumping into the icy water after food. In fact, they're tireless swimmers, frequently seen dog-paddling at six miles (10km) per hour. They've occasionally been found as far as 50 miles (80.5km) out at sea! Fat and two layers of dense, oily fur (very complex, with a hollow core) help them stay afloat.

IF IT'S MONDAY IT MUST BE WASHDAY

Have you ever watched the industrious RACCOON as it cleverly manipulates pieces of food with its oh-so-agile fingers (right up there with the chimpanzee!)? This "masked" mammal was originally named *Ursus lotor*, or the "washing bear," because of its reputation for washing its food before eating it. People who have watched claim they've never seen a wild raccoon wash its food. Scientists finally discovered that what the raccoons were really doing was dunking and then retrieving the food—it only looked like washing. Animal food got dipped more often than plant food. Strangely enough, the dirtiest food—earthworms—got dunked the least! But the animals in the study were in captivity, where all food comes from the land—and from a human provider. In the wild, the raccoon's food usually comes from the water—fish, snails, things like that. So eventually the experimenters realized that the raccoons were acting out catching water-food, *not* washing land-food. Your tame tabby will do the same thing if you give it a toy mouse. It will pounce on the pretend prey, throwing it around, imitating a real hunt!

Brian Parker, TOM STACK AND ASSOCIATES

RACCOON   *Procyon lotor*

IT'S TRUE THAT . . .

As you go north the raccoons get larger!

48

DON'T LET THE BLOOD GO TO YOUR HEAD

Probably one of the most bizarre mammals alive today is the South American TWO-TOED SLOTH. Sloths spend nearly all day—and all night, since they're really nocturnal—hanging upside down. Time to eat—upside down. Sleep—upside down. Mate, have a baby, feed the kid—all upside down! They use their long curved claws like hooks, to hang from. Because of these claws, they can't walk or even stand on their feet! Don't they get a headache looking at everything upside down? No need to worry—the sloth can turn its head 270°, so it can actually look at you right-side up if it feels like it. This turned-over tree dweller blends right in with the rainforest foliage. Algae growing along its hair strands tints the sloth—green in rainy weather, yellow during drought. From the ground this slow-moving creature with a *spurt* speed of 10 feet (3m) a minute seems almost motionless and it looks like just a bunch of leaves to a hungry jaguar!

BY THE WAY . . .

Two-toed sloths really have three toes on their back feet.

YOU NEVER GUESSED . . .

Three kinds of moths, certain beetles and many mites call the sloth's colorful fur home!

Gary Milburn, TOM STACK AND ASSOCIATES

TWO-TOED SLOTH   *Choloepus didactylus*

49

SUPER SKIN!

Who's 3,500 pounds (1,589kg), five feet (2.2m) tall and a vegetarian? You guessed it—it's the rhino, of course! Rhinos are pachyderms (that means "thick skin"), just like elephants, and some legends even claim that their hides are bulletproof! Of course, that isn't true, even though an approaching rhino does look as unstoppable as an armor-plated tank! You'd have to be awfully nearsighted not to notice a rhino if it charged at you, but scientists wonder whether the rhino, with its little-piggy eyes set far to the sides of the head, actually sees *you*! In other words, no one is sure if the rhino is charging because it *does* see you or because it *doesn't* and is simply running towards an unusual sound! Of course, if you get run down, it probably doesn't really matter.

WOULD YOU BELIEVE . . .

**The rhino's horn is not like the elephant's tusk.
It's made of glued-together hair.**

BLACK RHINO   *Diceros bicornis*

RINGTAIL or CACOMISTLE   *Bassariscus astutus*

SECRET SUNBATHERS

Many animals are shy and do their best to stay away from humans. Naturally, many of them do their roaming around at night. We call them "nocturnal." Animals that are active during the day are called "diurnal." But even night-roaming creatures seem to like the sun. This 2½-foot (.8m) long relative of the raccoon called the RINGTAIL or CACOMISTLE (Mexican for "bush cat") loves to come out for a sunbath. You won't see the ringtale; it does its basking in the very tops of the trees, stretching out fully along a branch, as if lying on a chaise longue!

51

LLAMA   *Llama quanicoe*

BUILT FOR THE HEIGHTS

The South American LLAMA (pronounced "yama") is a great example of "adaptive radiation." It has the same ancestors as the camel, which lives on the other side of the world, and similar characteristics, such as the split lip, long, curved neck and lack of skin between thigh and body. But after three million years of separation, it has adapted to a totally different environment. While the camel lives in the low-altitude heat of the desert, the llama lives in the cool mountains, at heights up to 13,000 feet (3,965m).

The llama is especially well equipped for dealing with the oxygen-poor atmosphere of the mountains. It never gets light-headed in the thin air. That's because its red blood corpuscles are unique among mammals. They are elliptical (egg shaped), and are able to take in more oxygen and live longer—235 days compared to 100 days for human red blood corpuscles—than the usual round cell. The ability to utilize all available oxygen is very important to a high-altitude dweller!

NINE-BANDED ARMADILLO   *Dasypus novemcinctus*

A REAL LIVE TANK

The largest armadillo is the three-foot (.9m), 130-pound (59kg) ARMADILLO of eastern South America. It got its name from the Spanish word *armado*, which means "one that is armed." Most of the armor in animals is made of hair that is pressed together, like the horn of the rhinoceros. But not the armadillo! Its protection is made of small bony plates, each one covered by a layer of hard skin. This walking tank is covered from head to tail with these plates that form a turtle-shell shape. Only its tummy is bare. If a threatening bird or animal shows up, the armadillo pulls its legs in and wedges itself down firmly on the ground—or rolls up in a tight ball—secure in its suit of armor.

DID YOU KNOW . . .

The nine-banded armadillo can submerge in water for six minutes!

PLATYPUS *Ornithorhynchus anatinus*

STRANGEST CREATURE ON EARTH

There are only two kinds of egg-laying mammals in the world (monotremes) and the PLATYPUS is one of them (the other is the spiny anteater—page 14). Early visitors to Australia could not believe their eyes when they saw this odd-looking creature. Here was a water-dwelling, burrowing, warm-blooded, furry creature that gave milk but laid eggs and dredged up dinner with a duck-like bill. At a loss for a name, they simply called it a "duckbill."

Dave Watts, TOM STACK AND ASSOCIATES

ODD BUT TRUE . . .

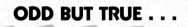

No fossil monotremes have ever been found!

Platypus ancestors have been around for over 135 million years, virtually unchanged! It seems that when Mother Nature created the duckbill she put some of her favorite animal structures together in a whole new way. Take the duck-like bill, for example: Its resemblance to a bird's beak is only skin deep. The platypus' dark grey bill is a moist, nerve-filled tool used to search muddy river bottoms for shrimp, worms and crayfish. It can sense the tiny electrical impulses given off by all living things, and this is a big help in finding its prey. When it resurfaces after a couple of minutes, its cheek pouches will be stuffed. Eventually, the platypus will eat the equivalent of 1,300 worms a day, while floating on the surface.

The platypus' front feet are webbed like an otter's, but the webbing extends beyond the end of the toes, creating a large paddle that can be pulled back, leaving the claws free for digging.

The strangest feature is the rattlesnake type of curved fang, called a poison spur (though with a less deadly venom), that is attached to the male's back legs. The spurs are used for defense, not for hunting. Sounds unbelievable, doesn't it!

A flat, muscular tail like that of a beaver brings up the end and is used as a swimming stabilizer and quick-dive assistant.

Now, imagine all this covered with fur (except the bill and webs)! Amazing.

TERROR OF THE TREETOPS

Not so! The giant GORILLA, as tall as a man and three times as heavy, eats only greens. His caveman display, pounding his chest, roaring, rearing up and throwing whatever's handy—usually just plants—is all show! Natives claim that gorillas will not attack a man who stands his ground, only those who turn and run away.

DID YOU KNOW . . .

Gorillas sleep in "nests" ten feet (3m) off the ground, curled up in the manner of humans!

Gorillas have an unusual animal "phobia." They're afraid of water! The mighty mammals are reluctant to cross the smallest streams. In the wild they don't even drink, preferring to get their moisture from their food—fruits and plants. The closest a gorilla willingly gets to water is soaking the fur on the back of its hand and sucking the water off!

HARD TO BELIEVE, BUT . . .

You'll know a gorilla is really mad when it sticks its tongue out at you!

GORILLA   *Gorilla gorilla*

Nancy Adams, TOM STACK AND ASSOCIATES

PACK RAT *Neotoma micropus*

NUTS TURNED TO GOLD

Can't find your favorite marble? Don't blame it on your kid brother—blame it on the PACK RAT. It looks like a big hamster about the length of your arm, though one-third of that is tail. These rodents like to live in the dry parts of North America. They also like to hoard "things"—especially bright, shiny or colored things—and frequently filch anything that catches their eye and is small enough to carry!

But, in their way, they're honest souls, usually leaving something behind in exchange.

One day some lucky miners discovered that the local pack rats had made off with a container's worth of shiny metal nuts, one by one, only to replace them with dull unprocessed gold nuggets! Now, that's a good deal!

What do the rats do with their treasures? Just like people, they build homes with "rooms" to house their possessions.

KOALA   *Phasolarctos cinereus*

IT'S A DANGEROUS WORLD

A living "teddy bear" and Australia's most beloved tree-dweller, the KOALA is usually a toy-sized two feet (.6m) tall and 33 pounds (15kg), covered by thick, soft, grey fur. It seldom ventures to the ground, and when it does, it's only to eat a bite of dirt (helps in digestion) or to scamper from one tree to another. The koala probably took to the trees in pursuit of its favorite and only food—the leaves of the eucalyptus tree, also called gum or tallow.

The koala's food fetish is specialization carried to a dangerous degree! Eucalyptus only grows in a slim slice of eastern Australia—the most heavily populated by humans—so that is where the koala must live. Out of the hundreds of varieties of eucalyptus, the koala eats only 12. Of these 12, each koala prefers five and has a single favorite. Add to this the fact that at certain times of the year food trees produce an overabundance of prussic acid, a deadly poison (¼ of a pound/.11kg would kill a sheep)! Fortunately, the poison builds up in different trees at different times.

At night, koalas climb to the top of the trees to eat the tender new shoots. If there aren't enough new shoots to make up the koala's 2¼-pound (1kg) daily requirement, they have to eat tough, older leaves with strong-smelling oils that are hard to digest. The good news is that these oils act like bug repellant, so koalas are lice- and vermin-free. The bad news is that koalas smell like super-strong cough drops!

BY THE WAY . . .

Koalas have a dozen names, among them: bangaroo, koolewong, karbor, narnagoon and most interestingly, New Holland sloth!

Life for the koala is a constant gamble. Because of its very limited diet, its survival could be in grave danger. If anything should happen to the eucalyptus trees, koalas might simply vanish from the face of the earth!

Brian Parker, TOM STACK AND ASSOCIATES

WOULD YOU BELIEVE . . .

Humans snore lying down, HORSES standing up!

Young DEER and OTTER play "Hide and Seek," but not together!

In a herd of WILDEBEESTS, all the pregnant females get together for a mass birth event!

BEAVERS' teeth are orange!

You'd need twelve GOATS to replace the milk from one COW!

Pampered CATS in ancient Egypt wore earrings!

In the long run, DOG teams are faster than HORSES!

SEA LIONS actually walk on their flippers!

All MAMMALS crave salt!

Exchanging a kiss is how PRAIRIE DOGS recognize accepted members of their community!

PRAIRIE DOG   *Cynomys ludovicianus*

Diana Stratton, TOM STACK AND ASSOCIATES

AND BESIDES THAT, DID YOU KNOW . . .

PORCUPINE   *Erethizon dorsatum*

CATS use their whiskers to feel their way in the dark!

The bacteria from the 50 million skin cells humans lose every day are what a tracking BLOODHOUND smells!

Easter Island is home to 4,000 wild HORSES and only 3,000 humans!

Contented BRAHMA BULLS purr!

The SPRINGBOK ANTELOPE can bounce straight up 12 feet (3.6m) in the air!

Bee-sting and snake-bite can't penetrate the HONEY BADGER's skin!

Cat connoisseurs claim the BOBCAT is braver than the LYNX!

In a dive, BEAVERS can swim about half a mile (800m) on one gulp of air!

PORCUPINES have as many as 30,000 spiny quills!

ASIAN LION *Panthera leo*

GALAGO (Bush baby) *Euoticus elegantulus*

MAMMALS EVERYWHERE

Mammals are everywhere—living on the ground like anteaters and antelopes; under it like rodents and rabbits; over it like bats; and even in the water like whales and walruses. Each one of these 4,008 species has something unique about its home, personality or lifestyle. This special something has developed so that the animal may survive successfully in the place, way or style that it does. Sometimes the trait may seem useless or bizarre, until we look further to see why it has developed.

Did the animal need flat teeth to grind plants or sharp teeth to bite meat? Colorful markings to attract attention or drab coloring to hide itself? Sharp claws to fight or strong legs for flight? Long thin arms to swing through the trees or flippers to swim through the sea?

For every obvious advantage, there are curiosities that even scientists cannot figure out. Why do you suppose that only the male platypus has a poison-spur, and why in such an out-of-the-way place as his back leg? Was there an ancient adversary, now extinct, that he had to defend against?

There is a lot left to find out. The more fascinating facts we learn about the mind-blowing mammals, the more questions we have. Perhaps one day you'll be the one to uncover the answers to some of nature's most tantalizing riddles!

Glossary

aborigine.   A member of the original race that lived in a region, such as an Aborigine of Australia.

algae.   Any of a large group of plants that contain chlorophyll but are not divided into roots, stems and leaves.

echolocation.   The sonar-like system used by dolphins, bats and other animals to detect objects around them through high-pitched sounds that are reflected and returned to the sensory organ of the sender.

fossil.   A trace or print of the remains of a plant or animal of a past age, preserved in earth or rock.

Ice Age.   The Pleistocene epoch (two million to 10,000 years ago), when sheet ice moved across the northern hemisphere.

lichen.   An organism composed of both algae and fungus cells that are usually dependent on each other for survival. Resembling a moss, lichen has no leaves, stems or roots.

lobe.   A rounded part of an ear. The projection at the end of an elephant's trunk.

mammoth.   A very large, hairy, extinct elephant with tusks that curved upwards.

monotreme.   An egg-laying mammal of the order Monotremata, now restricted to New Guinea and Australia—platypus and spiny anteater (echidna).

marsupial.   A mammal that gives birth to tiny, poorly developed young, which it carries in a pouch on the mother's stomach (kangaroo, opossum, and so on).

metabolism.   The process by which a living being uses food to obtain energy, build tissue and dispose of waste.

placental.   A mammal that retains the fertilized egg in its body until an advanced state of development is reached.

predator.   An animal that lives mostly by killing and eating other animals.

prehensile.   Able to grasp things, the way a monkey's tail can.

radar.   A radio device for detecting the position of things in the distance and the direction of moving objects.

sonar.   A method for detecting and locating things underwater by echolocation.

Jeff Foott, TOM STACK AND ASSOCIATES

POLAR BEAR
Ursus maritimus

Index